QUARK
H
N
H
Na
ENERGY = 1.5
Sector A
Sector B

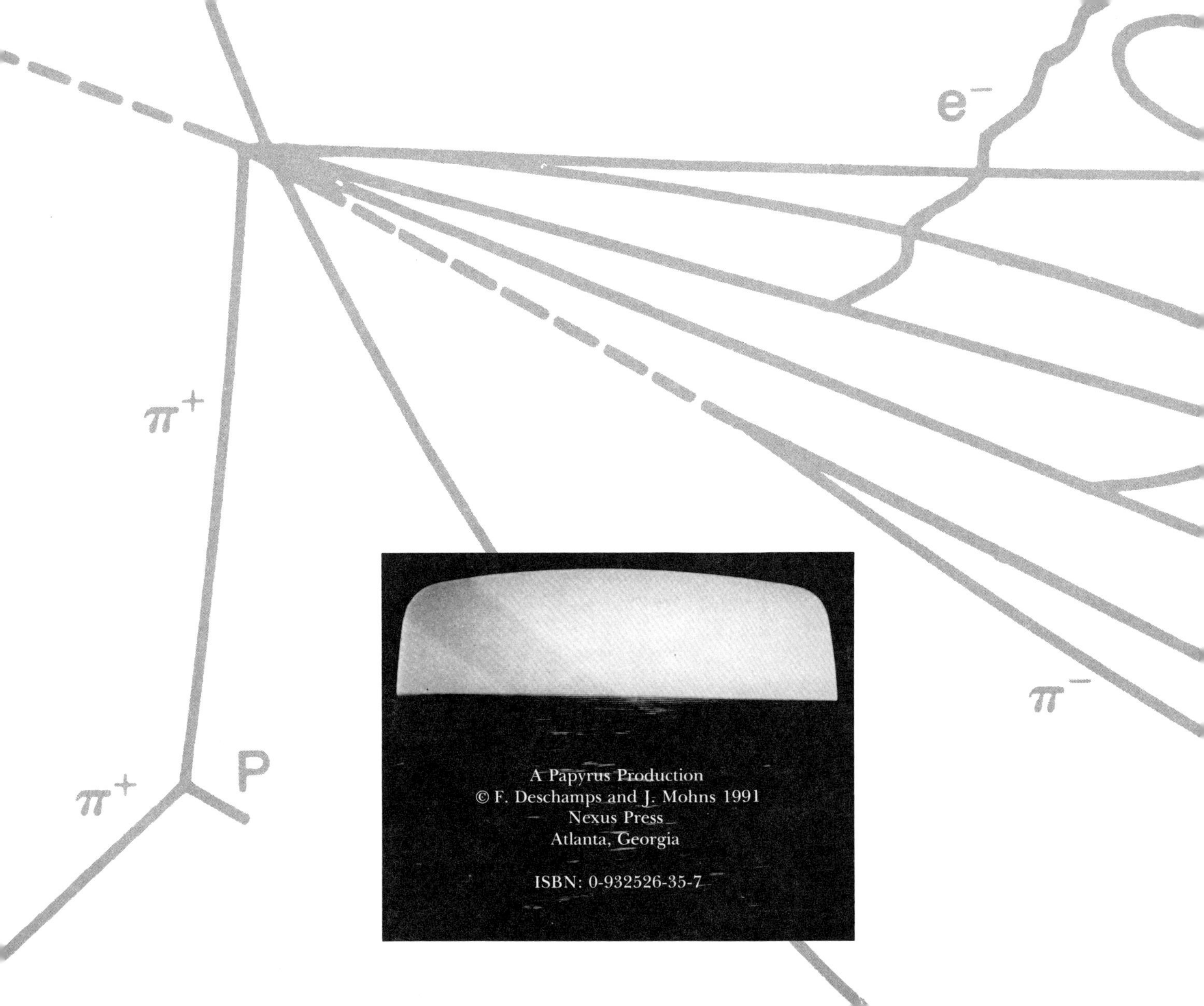

A Papyrus Production

Nexus Press
Atlanta, Georgia

ISBN: 0-932526-35-7

PARTICLE THEORY

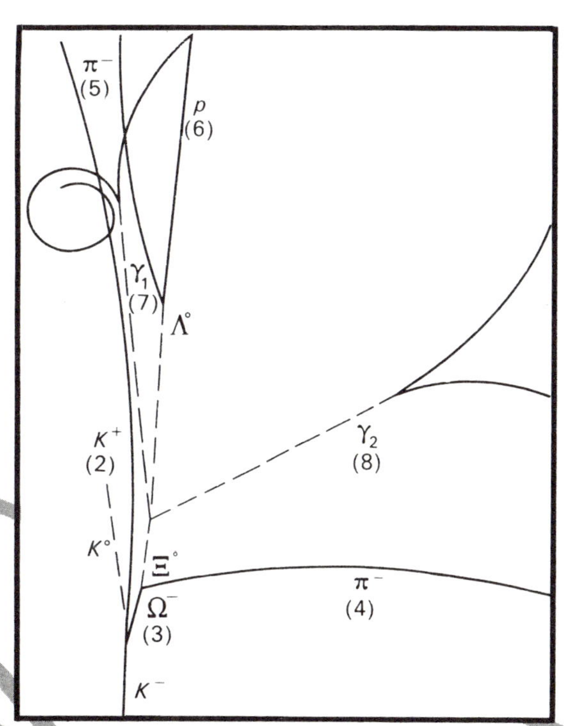

F. DESCHAMPS and J. MOHNS

INTRODUCTION

Particles are now recognized as the universal building blocks of all systems, be they chemical, biological, cultural, or historical. Scientists and scholars try to understand the world in these terms by proposing various particle theories. This book describes many of these theories in simple language, making this exciting area of study easily accessible to the layman. The text is divided into seven units. Each unit concludes with challenging exercises to help our readers achieve a deeper understanding of Particle Theory.

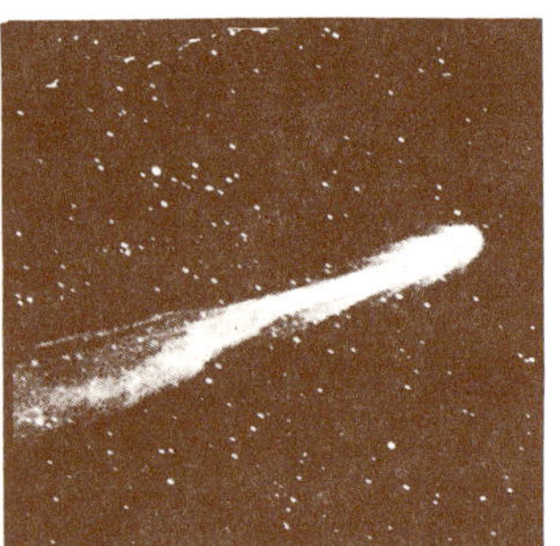

TABLE OF CONTENTS

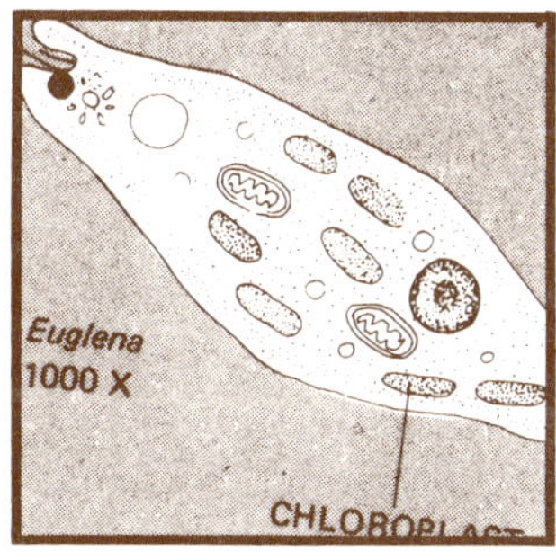

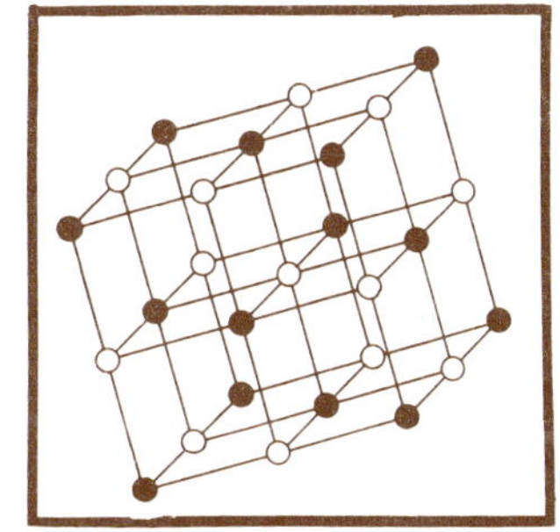
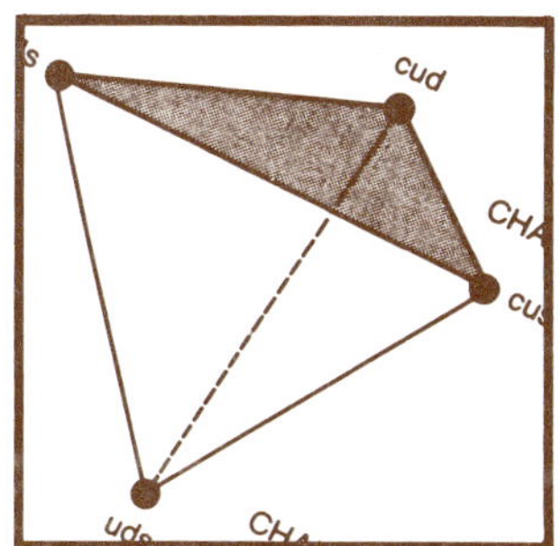

UNIT 1: GENERAL PARTICLE THEORY

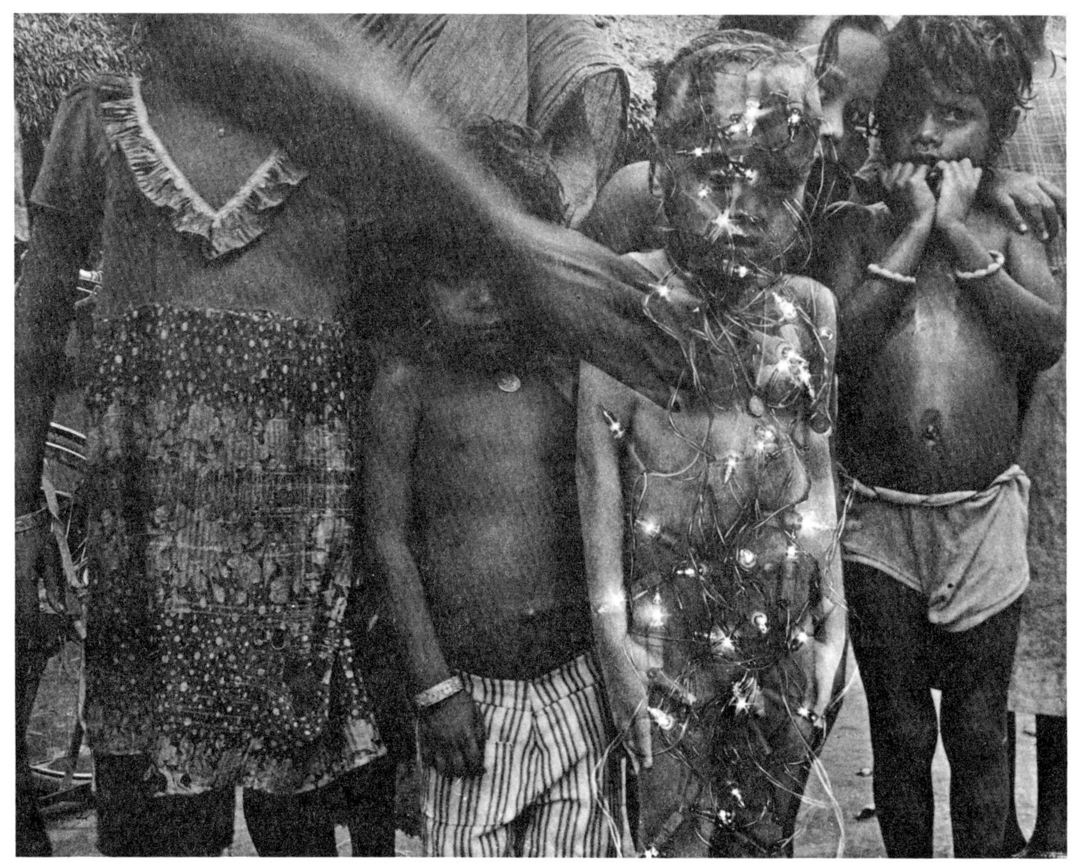

The soul does not reside in any one part of the body but is a diffuse soup of luminous particles spread evenly throughout the body.

A particle is a solitary unit of a physical, political, cultural, or biological system. For instance, a human being is a particle in a political system, as bugs are particles in an ecosystem. But those same beings are composed of smaller particles that are the cells of their biological systems. Images can also be viewed as particles. For example, images make up a cultural system, and a single culture is a unit of a global system.

The concept that Life is composed of particles is an old idea with origins in Ancient Greece. The Roman poet Lucretius even wrote a long poem in which he called particles "the seeds of things" (*semina rerum*).

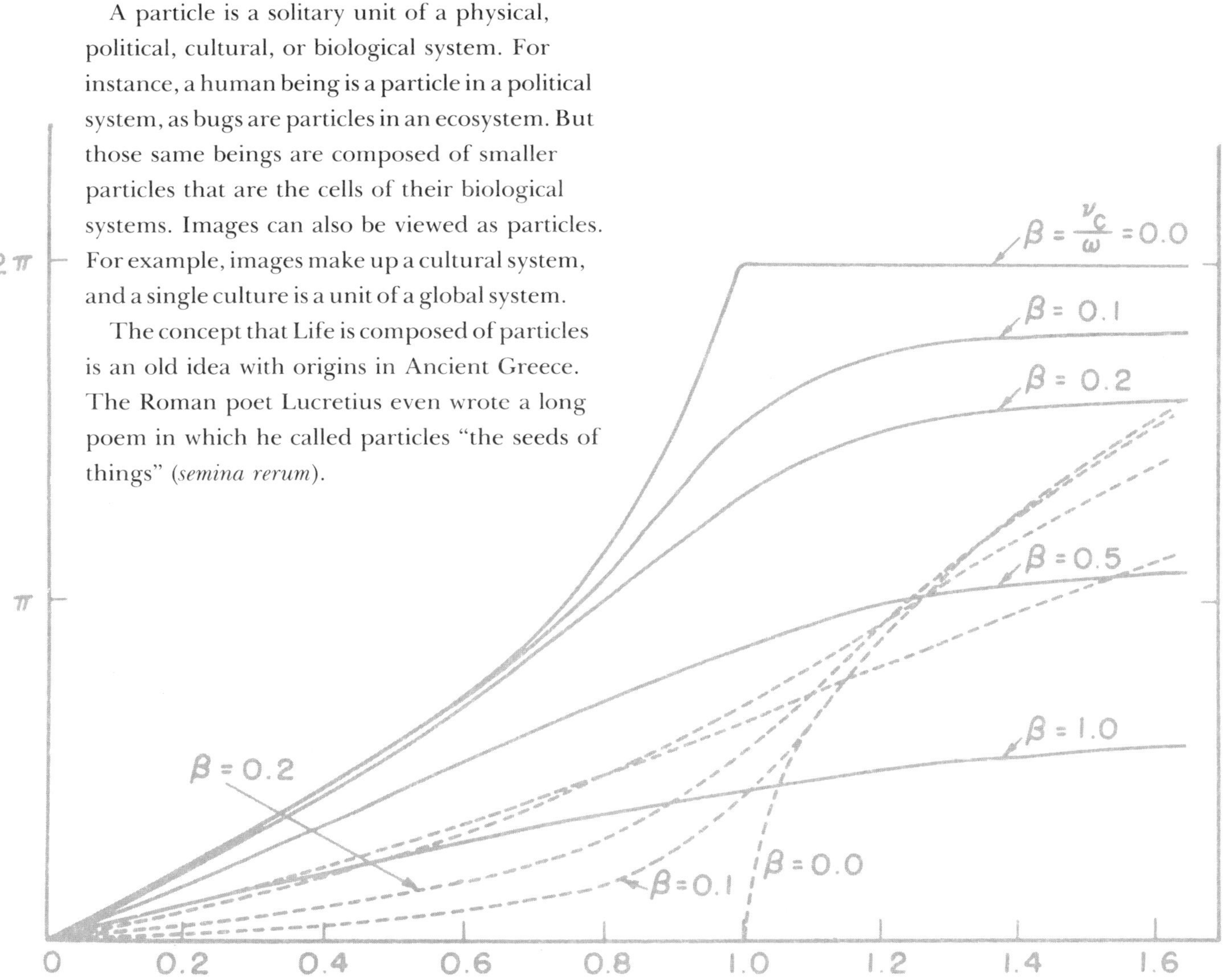

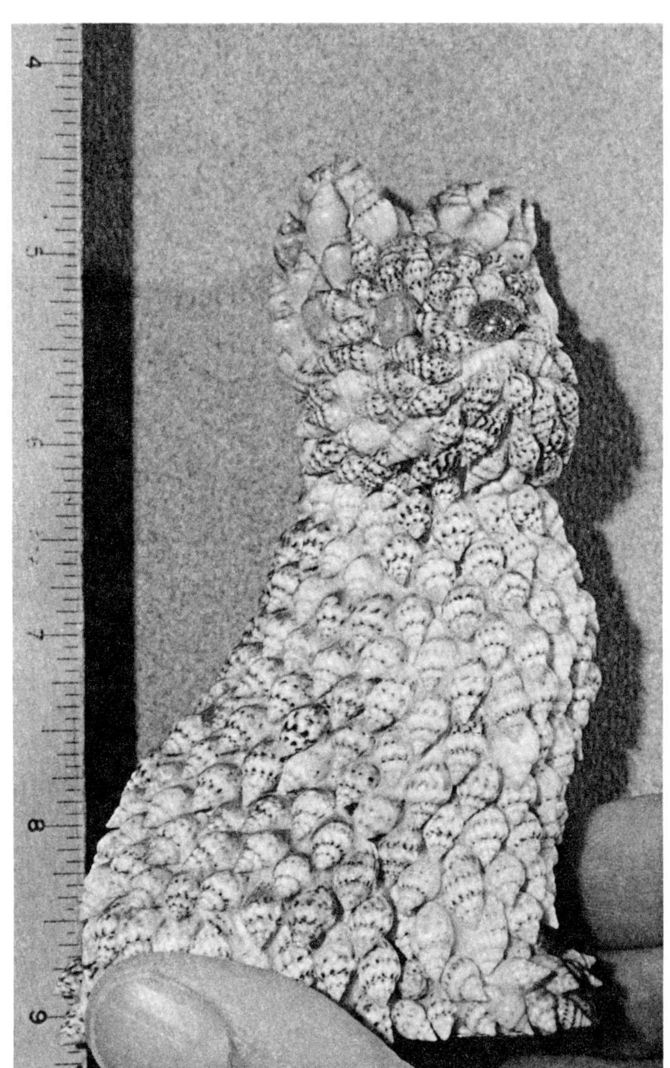

Dr. Hermann Wolfe was one of the first scientists to work with constructive Particle Theory back in 1879. He based his theory on study of the ancient sacrificial statues made by the Druids, which were composed of caged humans. In Wolfe's principal experiment, he was able to create a small dog made entirely out of seashells.

GIGANTIC DRUIDICAL IDOL.

μ+
e+

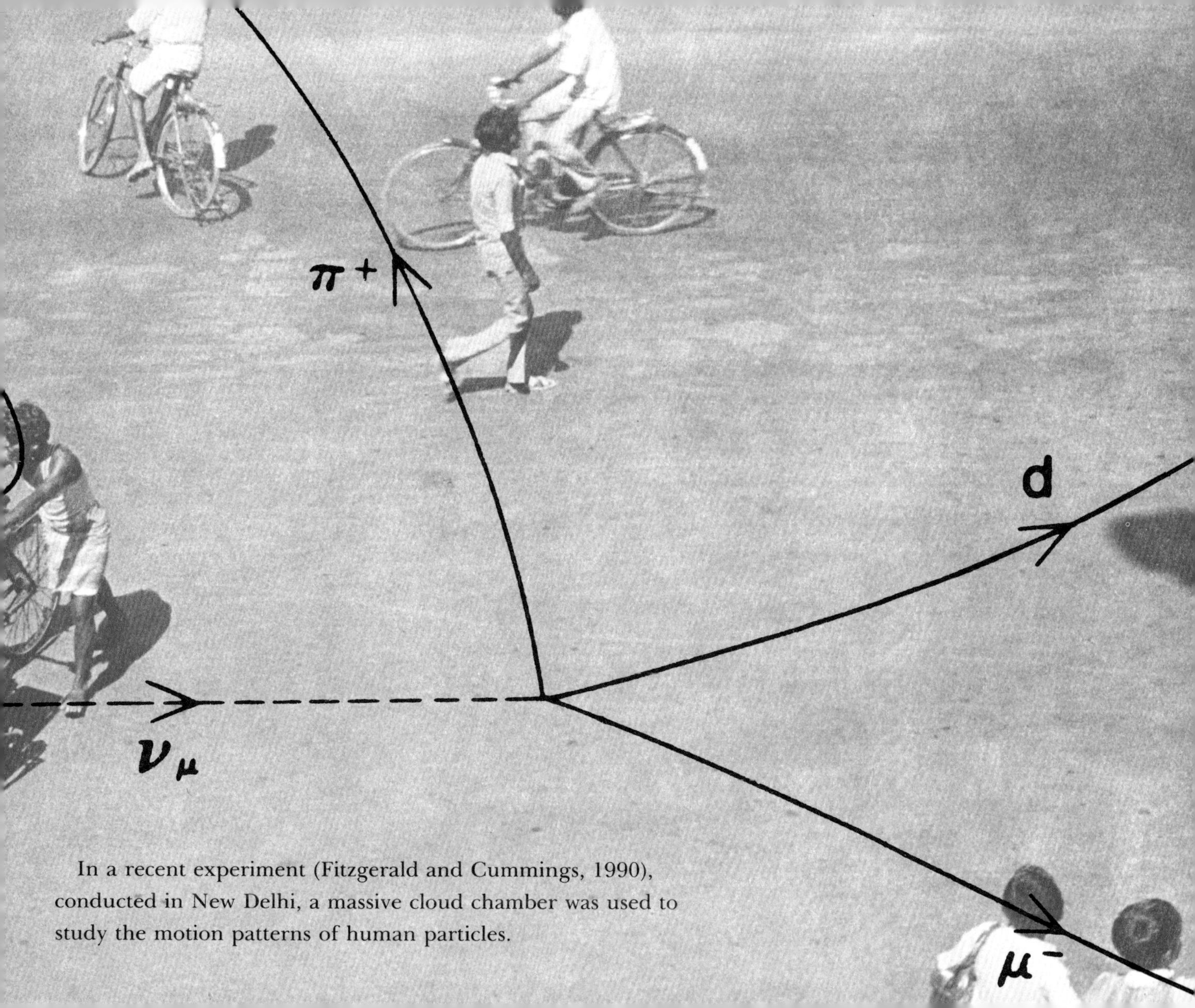

In a recent experiment (Fitzgerald and Cummings, 1990), conducted in New Delhi, a massive cloud chamber was used to study the motion patterns of human particles.

UNIT 1: EXERCISES

1. What is a particle? Are you a particle?
2. Who wrote a long poem about particles?
3. Do you live in a cloud chamber?
4. List the systems of which you are a particle.

UNIT 2: THE PARTICLE THEORY OF REPRESENTATION

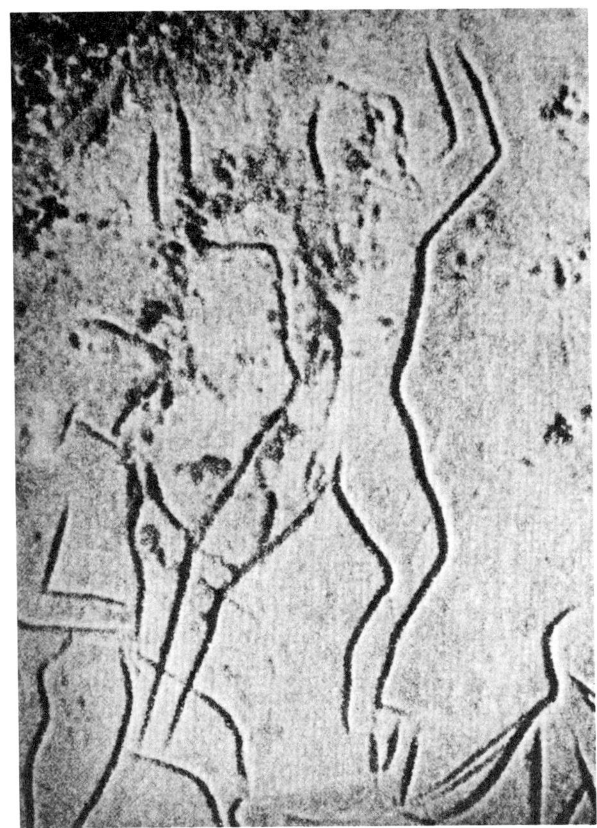

10,000 B.C.

1873

1934

1987

Throughout history, representation has always been achieved by the use of particles.

In recent times, technology has allowed human beings to be visualized as dense accumulations of particles.

This representation of humans as dots or particles has led philosophers to perplexing problems concerning the nature of self and political behavior. For instance, the theory of democracy and majority rule essentially rests on the idea that dots accumulate in either one or another box. If there is just one dot more in Box A, then Box B loses—thus defining political truth.

Scientists have found that image particles are susceptible to the same ailments that can affect larger systems. In a study conducted at Cornell University (Gregg, 1989), imagery succumbed to a biological infestation. The small moth (shown on this page) is feeding on the halftone dots that construct the image. It is feared that such disease can affect not only the visual details but also the meaning of images.

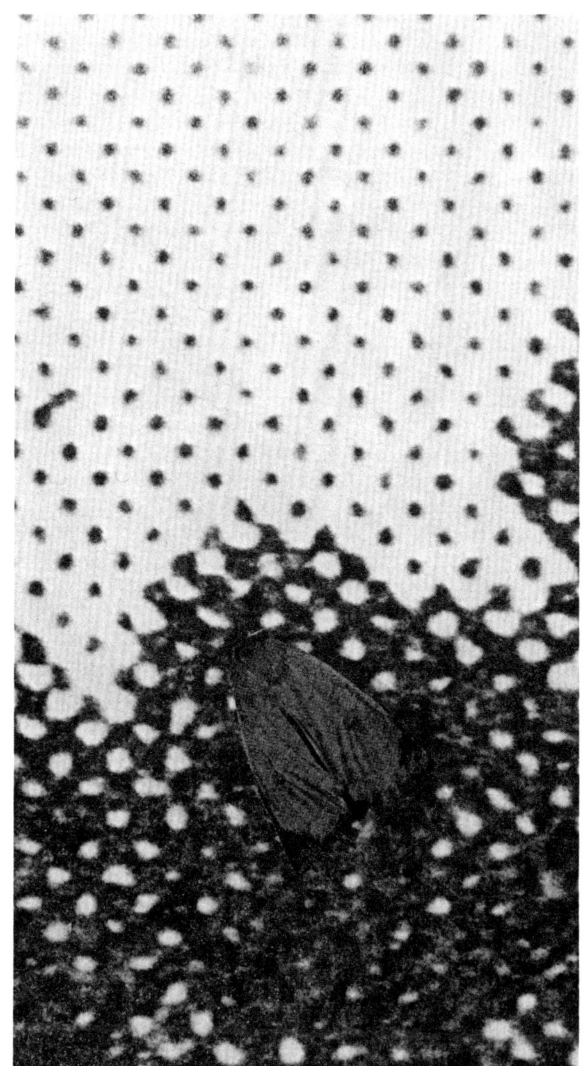

UNIT 2: EXERCISES

1. Do you like to look at images?
2. Are the best images made of the smallest particles?
3. Are images important?
4. Do you make decisions by looking at images?
5. What happens if the particles that make up an image become diseased?

UNIT 3: THE VIRAL THEORY OF KNOWLEDGE

The Viral Theory of Knowledge (Jeffers and Cheek, 1974) proposes that information and ideas spread like disease among humans. Certain knowledge spreads very quickly with little lasting effect, while other information can be deadly if not treated immediately. According to this theory, knowledge is a parasite that lives in the brain cells of human beings. It has been said that Western Civilization is a "viral infection."

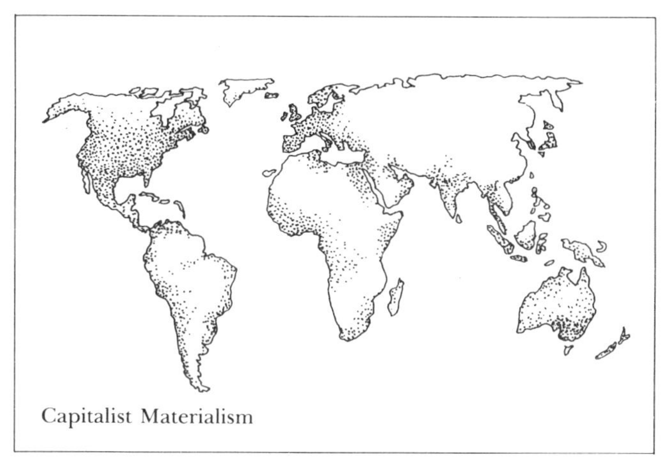
Capitalist Materialism

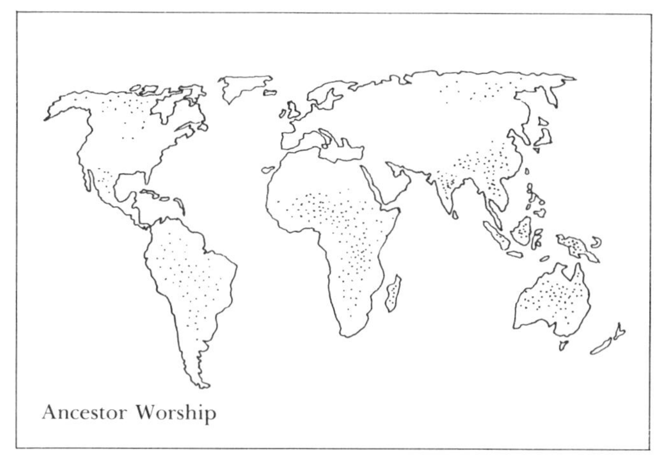
Ancestor Worship

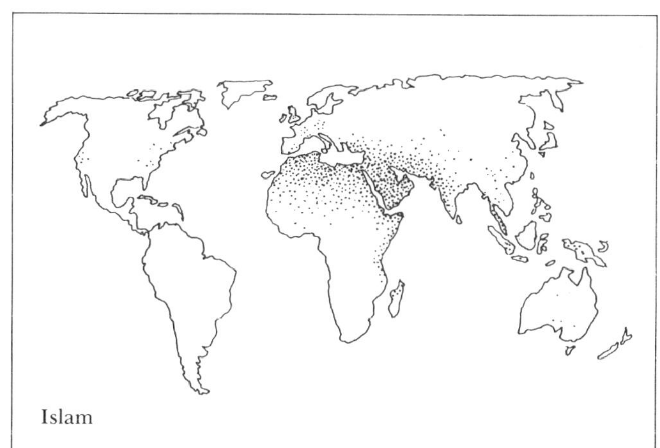
Islam

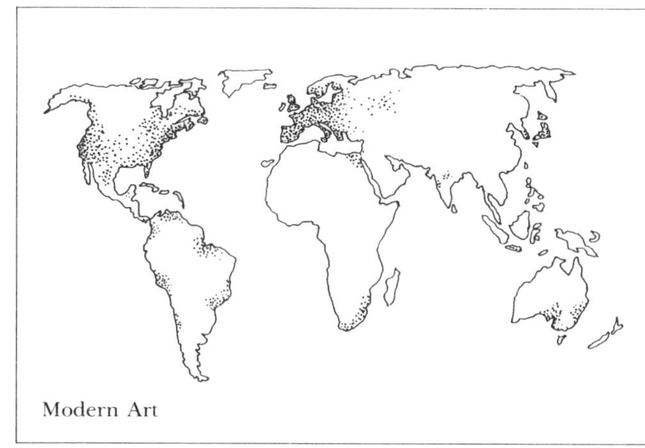
Modern Art

Le Clic
TUFF 35
ESTWING
120
110
100
90
80
70
60
50
40
30
20
10
0
10
20
30
40
Skil Twist

In addition, those viral ideas cause people to produce objects for various uses. These objects then become carriers and seeds of viral ideas.

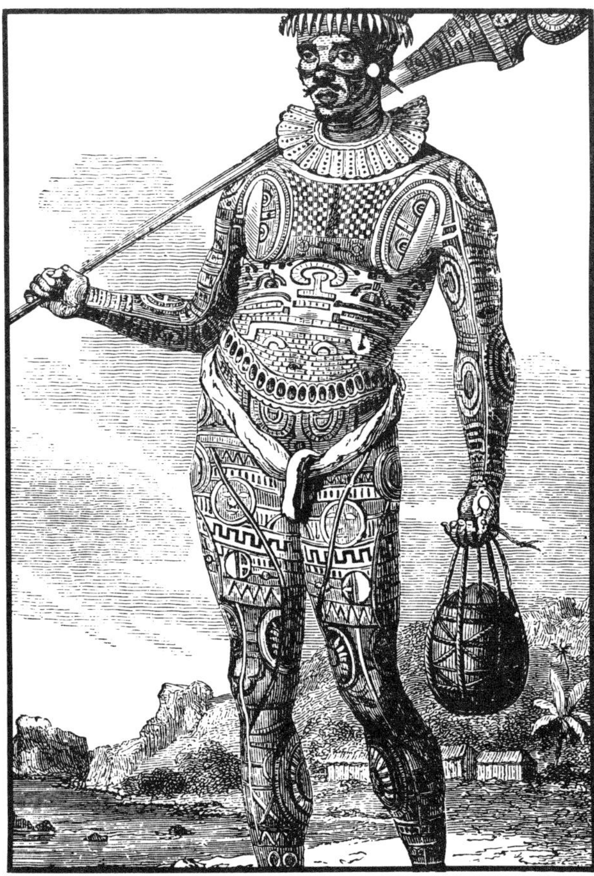

Tattooed savages show a variety of cultural viral afflictions.

The controversial product Photo Seeds is another striking example of the potential power of viral infestation. These "seeds" were marketed in the 1970's as a harmless novelty item. The product, however, was pulled off the market when it was discovered that unscrupulous manufacturers were putting dangerous images in the seeds.

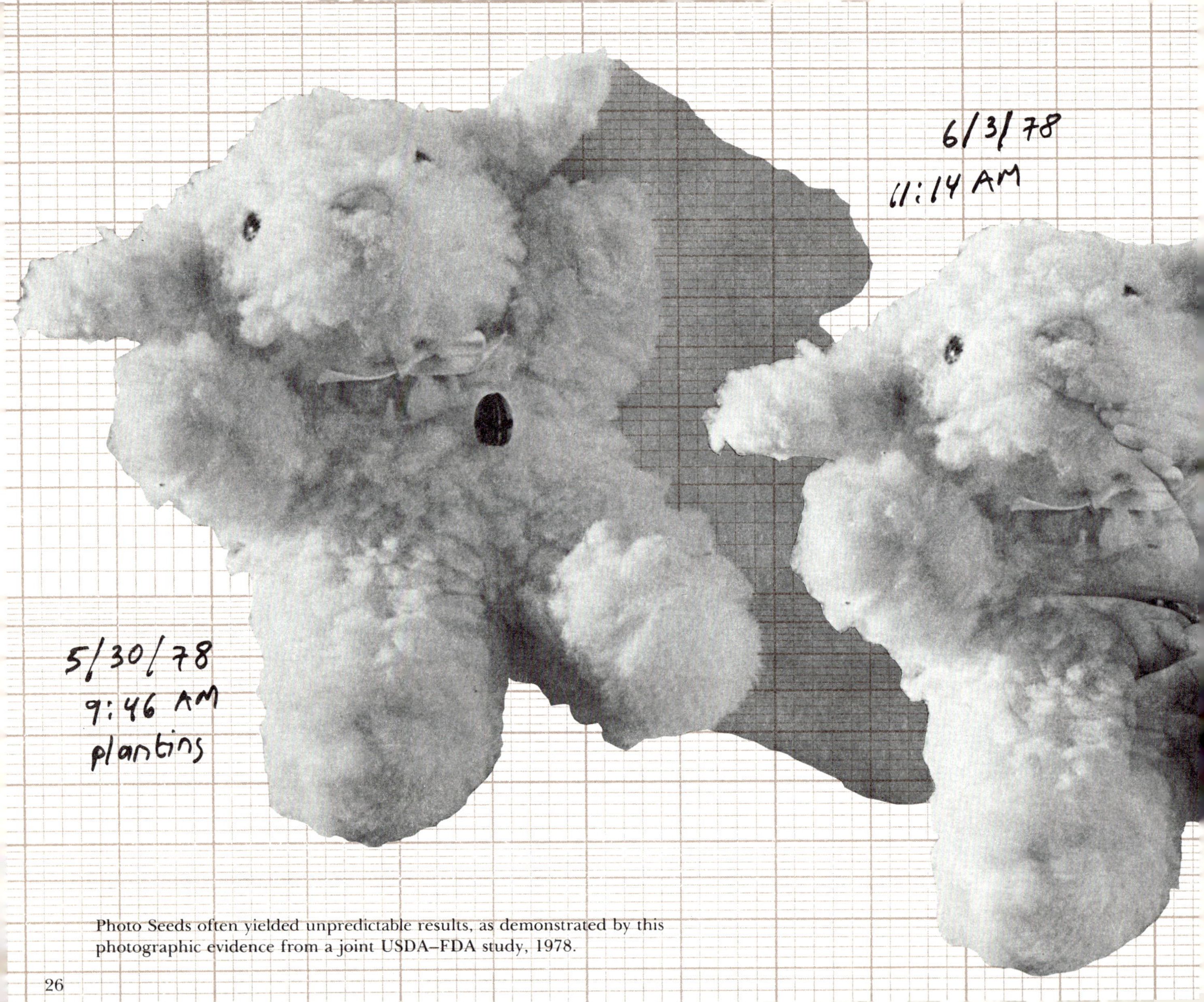

Photo Seeds often yielded unpredictable results, as demonstrated by this photographic evidence from a joint USDA–FDA study, 1978.

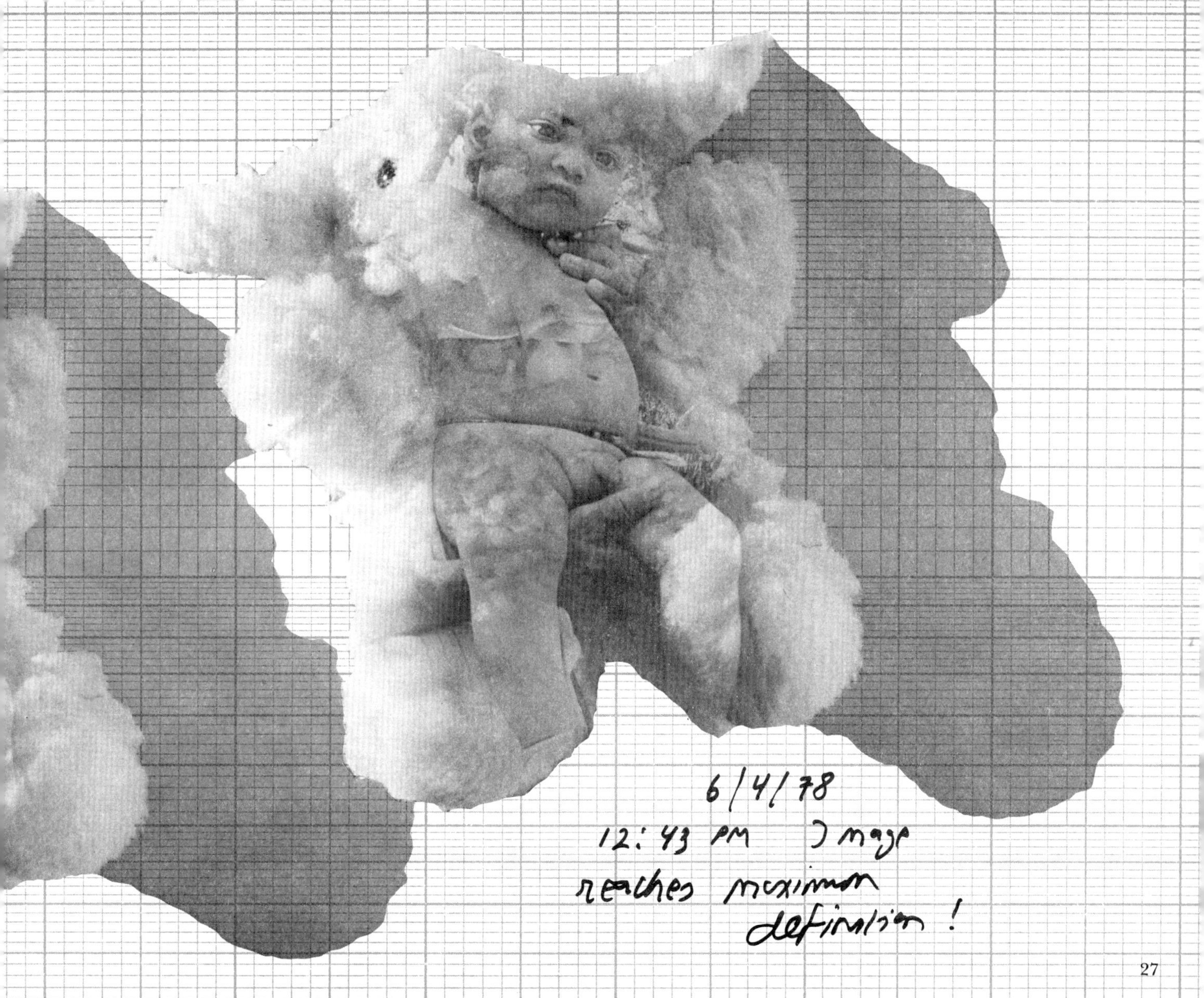
6/4/78
12:43 PM Image
reaches maximum
definition!

UNIT 3: EXERCISES

1. Does your brain have viruses? If so, name some.
2. Are some brain viruses good?
3. Why are Photo Seeds dangerous?
4. Where would you plant Photo Seeds?

UNIT 4: THE TOPOLOGY OF IMAGERY

The photograph of an object can become
the object of a photograph.

An image of a broken thing can be repaired by breaking the original image.

Images exist within a culture and, therefore, are particles of that culture. Culture can be viewed as a space populated by images. Since topology is a branch of mathematics that concerns itself with space, topological concepts can be applied to images in the cultural space. Initially, topologists investigated neutral images.

The Impossible quadrilateral
(Sept., 1981)

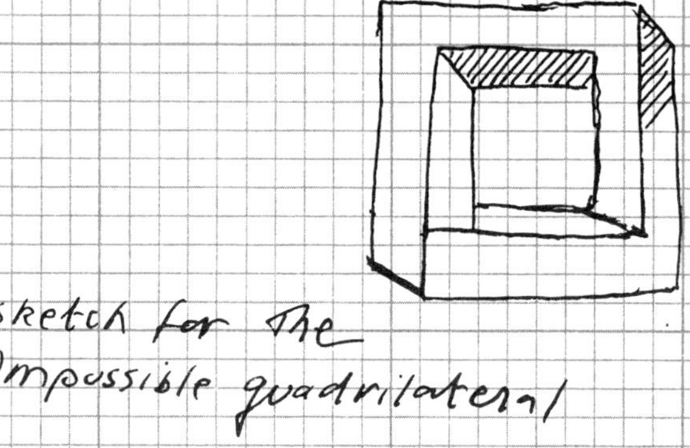

sketch for The
Impossible quadrilateral

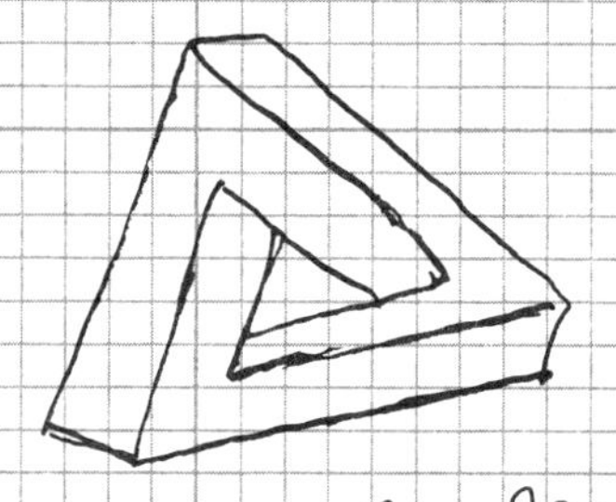

The Penrose Tribar
(sketch, 1978)

Then these topologists experimented with so-called impossible objects that have no referent or model in the real world. A major breakthrough occurred, which now allows photography of objects that cannot exist in normal three-dimensional space (Friedman and Hoyt, 1981).

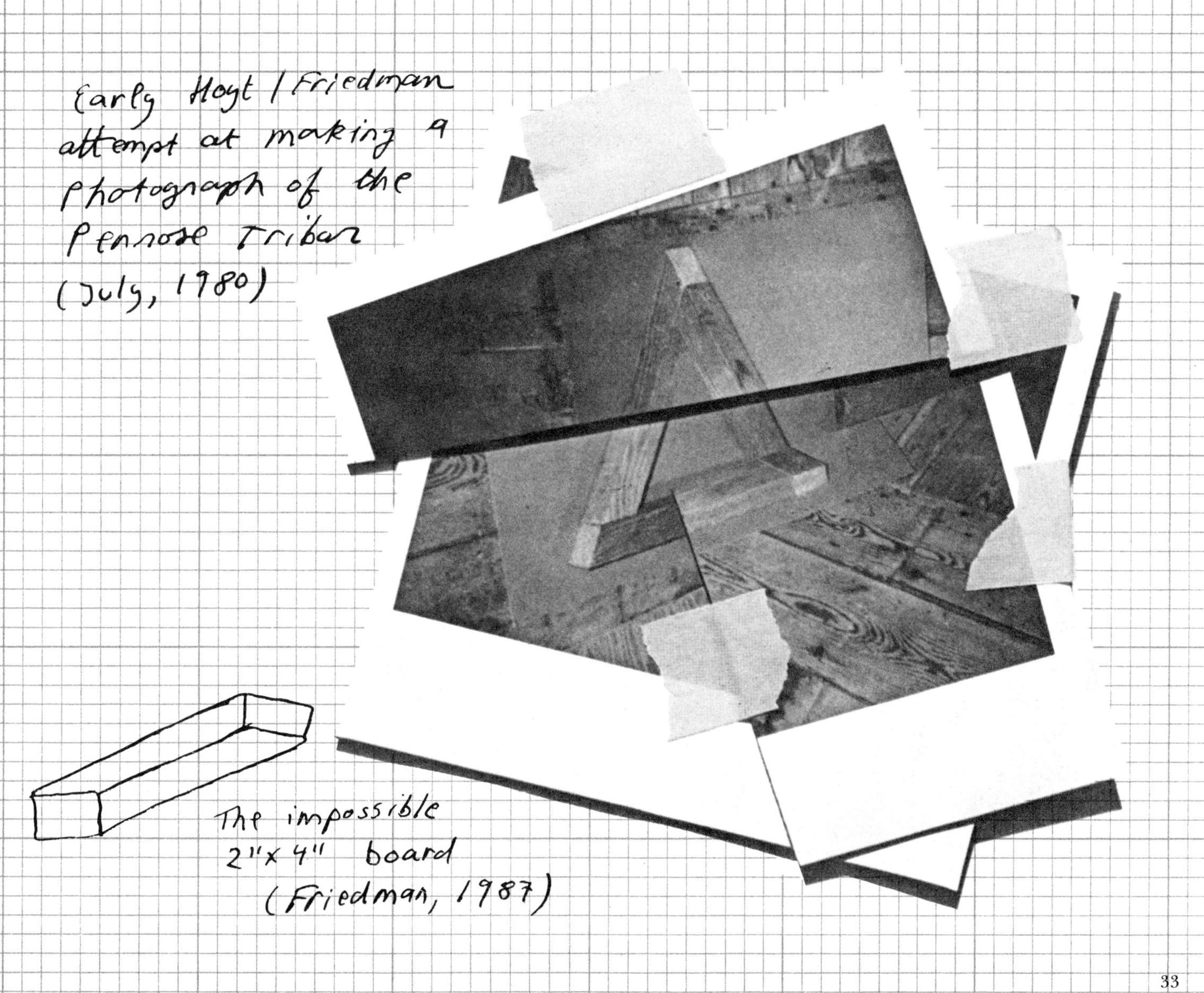
Early Hoyt / Friedman
attempt at making a
photograph of the
Penrose Tribar
(July, 1980)
The impossible
2"x 4" board
(Friedman, 1987)

Another new area of investigation involves the "arbitrary object." Scientists McCarney and Smith spilled a glass of water 2,671,341 times—until the spill matched a predetermined photograph of a bird.

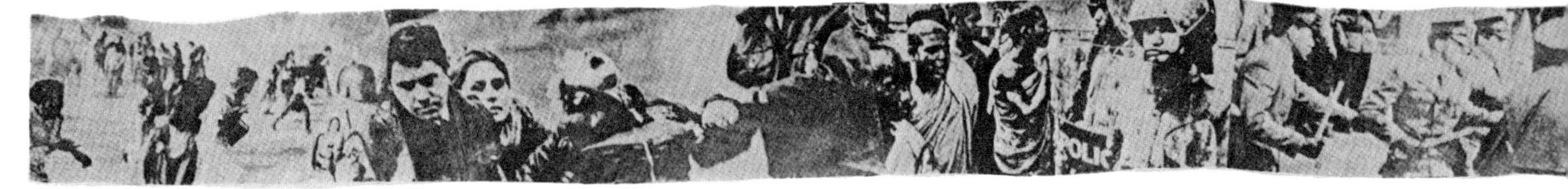

Inside Surface

A more practical application of image topology is exemplified by the political Möbius Strip. The Möbius Strip is a three-dimensional shape with only one side, unlike a piece of paper that has two sides. To understand how it is constructed, look at Figure 1. The inside surface has particles of people fighting in poor countries. The outside surface has particles of pretty things in rich countries. By cutting along line $\overline{AB}$, reversing one side and reassembling, a Möbius Strip is created. See Figure 2. Now the poor people fighting are on the same surface as the rich people with their pretty things.

Outside Surface

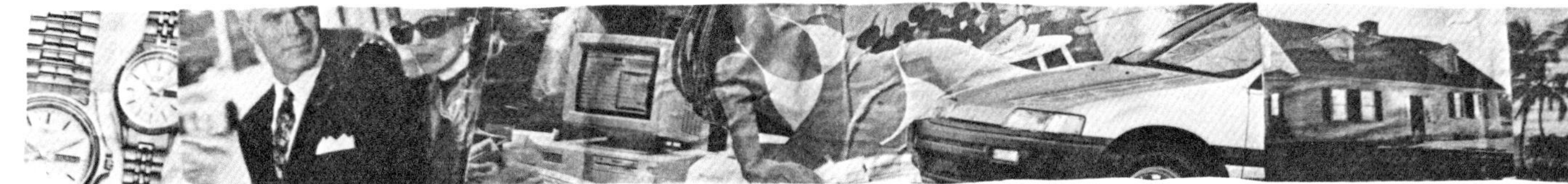

Figure 1. The Old Order; Two Surfaces

A

B

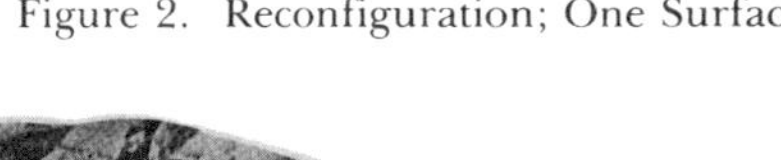

Figure 2. Reconfiguration; One Surface

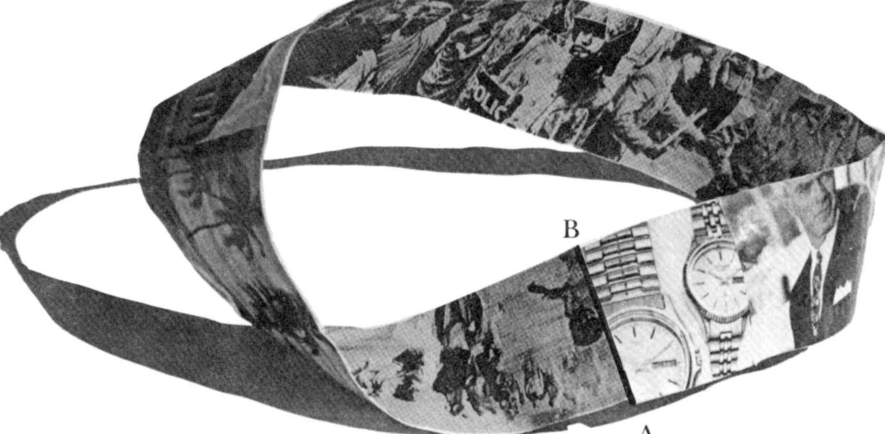

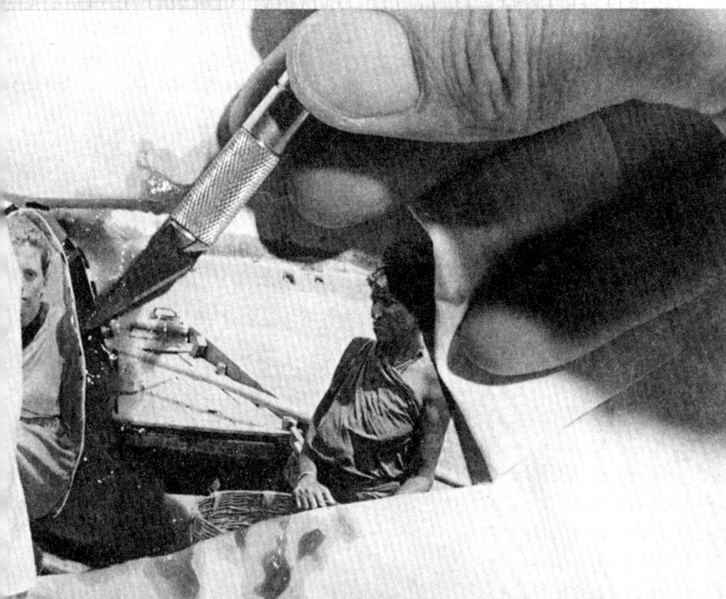

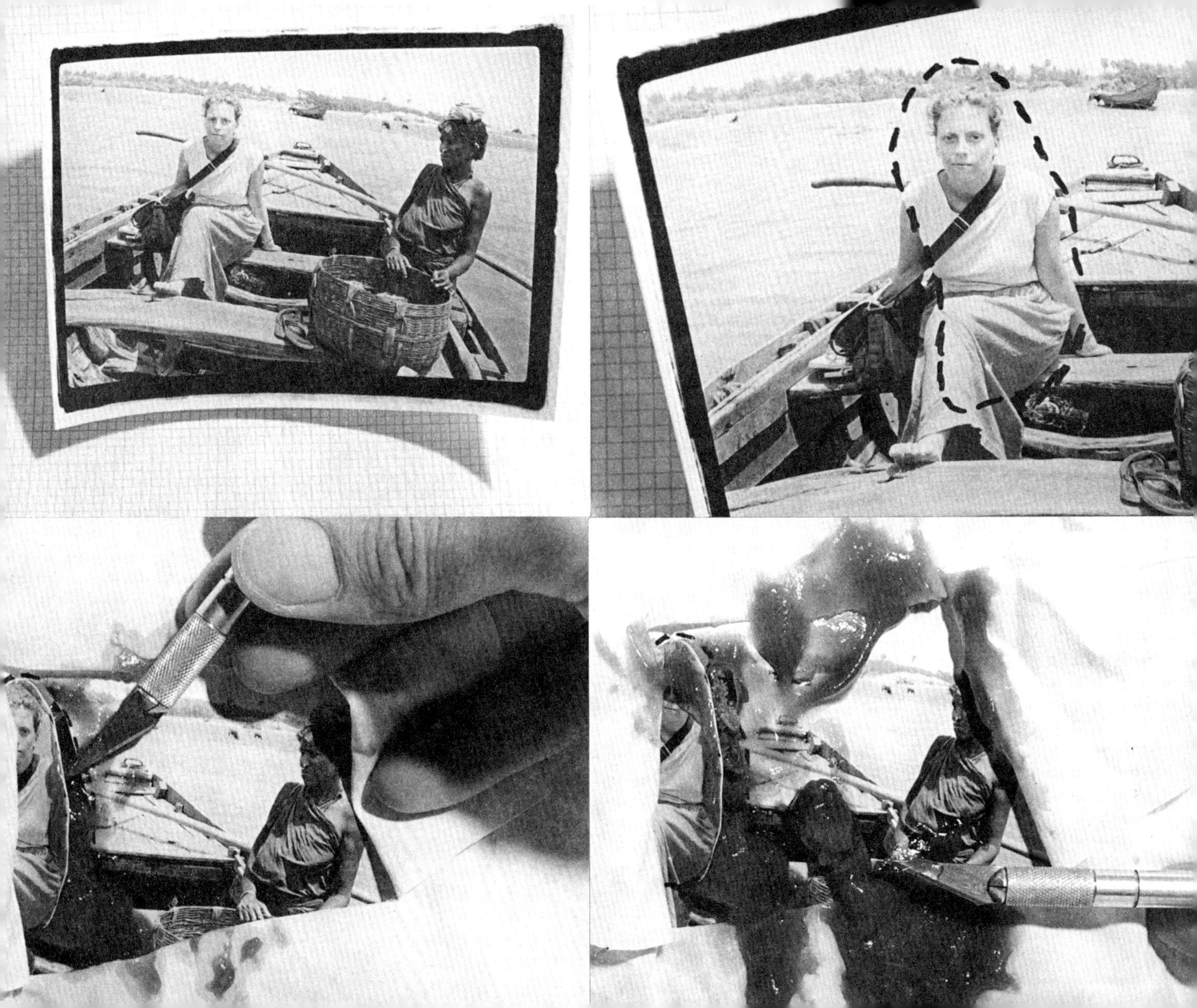

One of the most daring applications of topological procedures applied to imagery involves surgical excision. While the procedure can be somewhat bloody, it is safe and the stitches soon disappear, leaving no trace.

UNIT 4: EXERCISES

1. Do you like poor people? Are you rich?
2. Is a political Möbius Strip dangerous?
3. Which surface of the uncut cylinder are you on?
4. Why do topologists make impossible objects?
5. Is it safe to do surgery on images?

UNIT 5: *PARTICLE THEORY*: THE SOAP OPERA

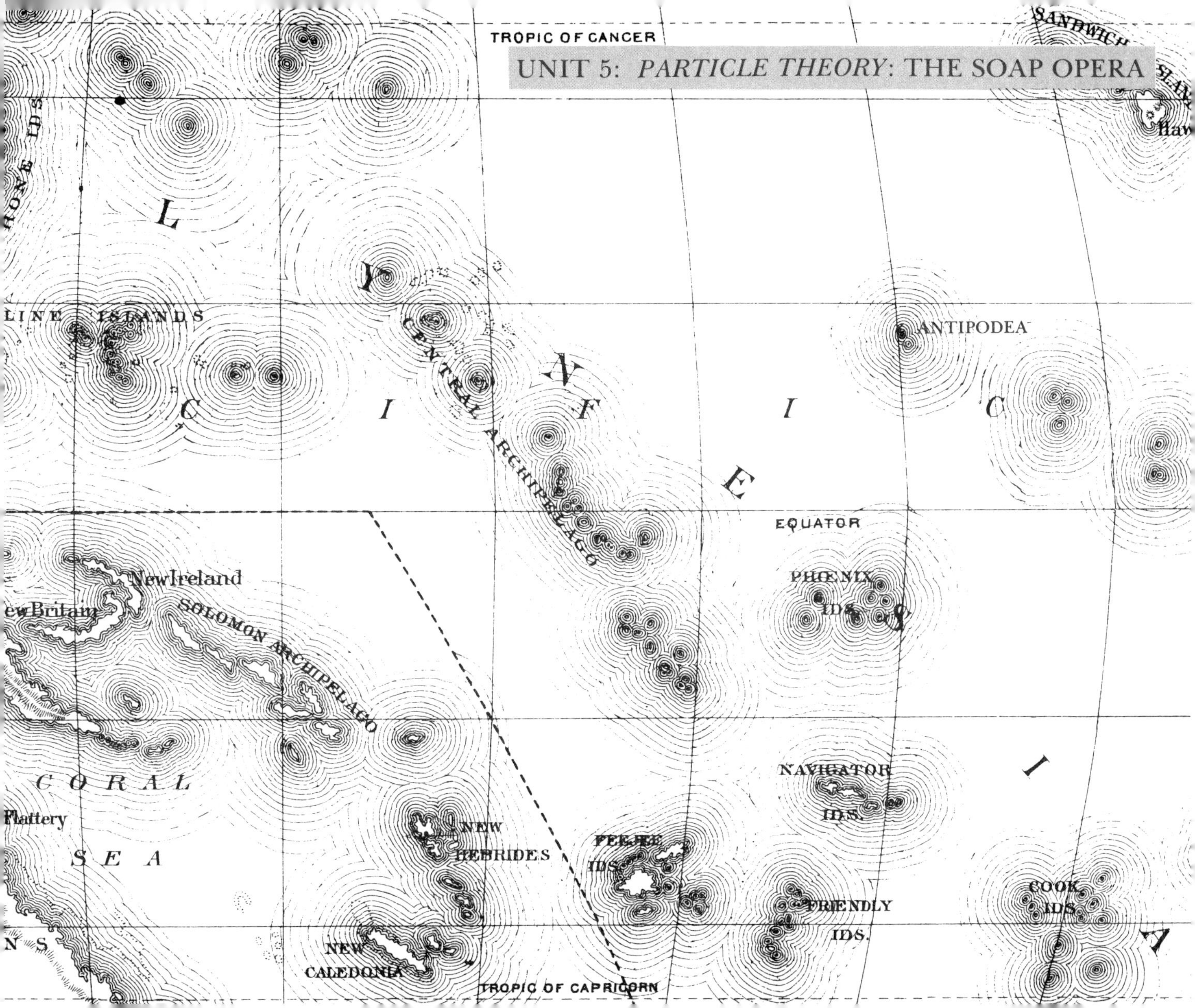

An interesting application of the preceding theoretical considerations is the strange case of the daytime soap opera *Particle Theory*. Writers asked themselves the question: "What if a Third World man became the star of a Western soap opera?"

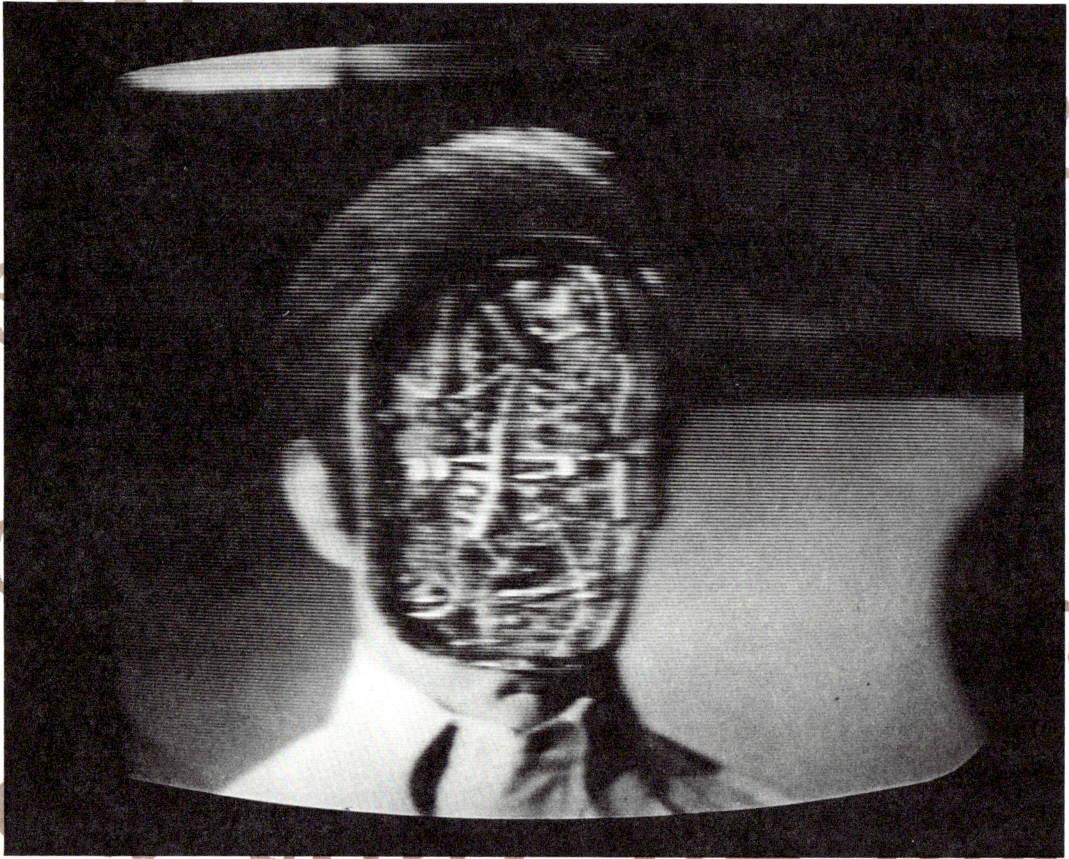

NOBLE SAVAG

aw a savage runnin

d as naked as when h

nintelligible *bêche-de*

d to be our servant

The name of the hero was Em Etah Uoy. In the series, he came to the United States to retrieve a sacred icon, the Amulet of Light, from a museum. His plan was to return it to his native island of Antipodea. Em Etah Uoy worked as a janitor in a particle lab and happened to fall in love with Pat Lamb, a typist at the museum.

The writers used authentic details from historical and anthropological studies to give the soap opera realistic touches. Em Etah Uoy became an enigmatic and challenging hero for the American public. His immense popularity could not have been anticipated. He was charming and witty one moment, then cruel and vicious the next, leaving viewers confused over his plight. At one point Em Etah Uoy and the writers had a celebrated exchange of letters that appeared in *Soap Opera Digest*.

Caribs torturing a prisoner, whose flesh they devour while he is still alive.

Dear Em,

You seem so calm and pure, as if nothing could hurt you. You flow easily in a crowd, in a foreign land.

We need privacy, a watch, a schedule, a calendar, an accountant, even a doctor to keep a record of our most intimate fluids--blood, urine....

You could live or die easily it seems. You are content to be fragile. This gives you a spark of intensity that we lack. You know how precious each moment of your life is. We are too busy measuring and protecting ours to notice its exuberance.

Beyond food, shelter, sex, we need security. Security is our ultimate vice. We need higher and higher doses to feel relaxed. Unfortunately, our obsession with security leaves us constantly insecure--constantly aware of the dangers.

We admire your adventure and your adventurousness. We don't understand the thing about the Amulet of Light, but perhaps we will as the story goes on. You could get friendly with Pat, but be careful because many people in America do not like interracial relationships.

Good Luck!

Sincerely Yours,

The Writers

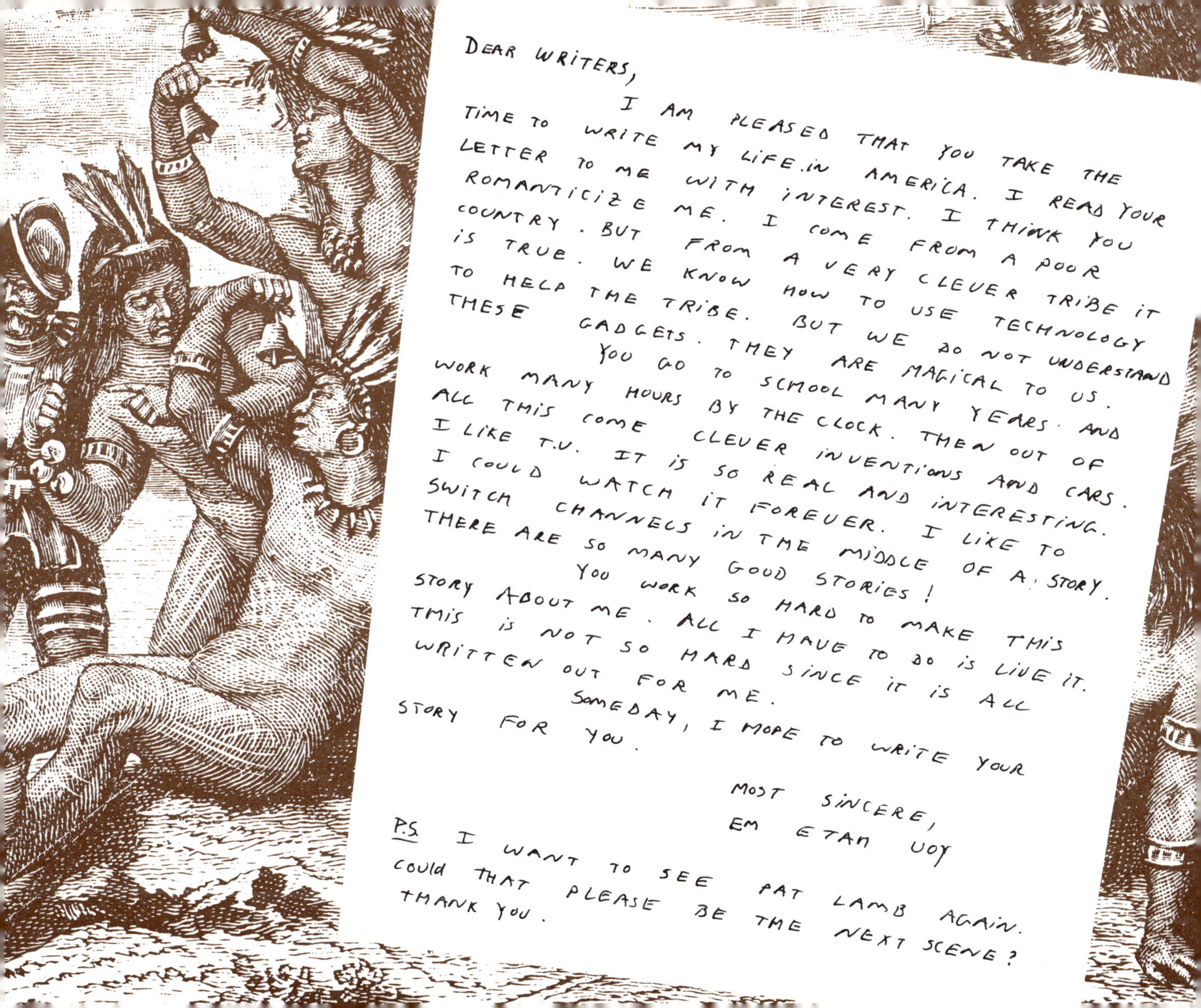

DEAR WRITERS,

I AM PLEASED THAT YOU TAKE THE TIME TO WRITE MY LIFE IN AMERICA. I READ YOUR LETTER TO ME WITH INTEREST. I THINK YOU ROMANTICIZE ME. I COME FROM A POOR COUNTRY. BUT FROM A VERY CLEVER TRIBE IT IS TRUE. WE KNOW HOW TO USE TECHNOLOGY TO HELP THE TRIBE. BUT WE DO NOT UNDERSTAND THESE GADGETS. THEY ARE MAGICAL TO US.

YOU GO TO SCHOOL MANY YEARS AND WORK MANY HOURS BY THE CLOCK. THEN OUT OF ALL THIS COME CLEVER INVENTIONS AND CARS. I LIKE T.V. IT IS SO REAL AND INTERESTING. I COULD WATCH IT FOREVER. I LIKE TO SWITCH CHANNELS IN THE MIDDLE OF A STORY. THERE ARE SO MANY GOOD STORIES!

YOU WORK SO HARD TO MAKE THIS STORY ABOUT ME. ALL I HAVE TO DO IS LIVE IT. THIS IS NOT SO HARD SINCE IT IS ALL WRITTEN OUT FOR ME.

SOMEDAY, I HOPE TO WRITE YOUR STORY FOR YOU.

MOST SINCERE,
EM ETAN UOY

P.S. I WANT TO SEE PAT LAMB AGAIN. COULD THAT PLEASE BE THE NEXT SCENE? THANK YOU.

VICTIM OF HEADHUNTERS

Eventually, interest in *Particle Theory* faded, and the writers frantically tried to wrap up loose ends before the series was cancelled. A new show about a giant talking carrot would be filling the same time slot. Although Em Etah Uoy had been able to retrieve the Amulet of Light, the writers were still left with the problem of concluding the story.

By their choice of ending, the writers would imply a certain viewpoint on culture and politics. They considered the following:

1. One day Em Etah Uoy sees a documentary on television about Antipodea, gets very homesick, comes to his senses, and returns home with the Amulet of Light. This shows that Primitive and Western cultures are separate but equal.

2. Em Etah Uoy deteriorates to the point of becoming a drug addict who loves gambling and pornography. This shows that Western culture is in decay, and that the old cultures are too weak to save it.

3. Em Etah Uoy stays in the United States, builds a life and a family with Pat, and becomes a model immigrant. He stays faithful to the Amulet of Light, which shows that a synthesis of cultures is possible and desirable.

4. The Amulet of Light comes alive, influencing Em Etah Uoy's life by talking to him and making him return home. This shows that Primitive cultures far surpass our pedantic, Western, scientific thought.

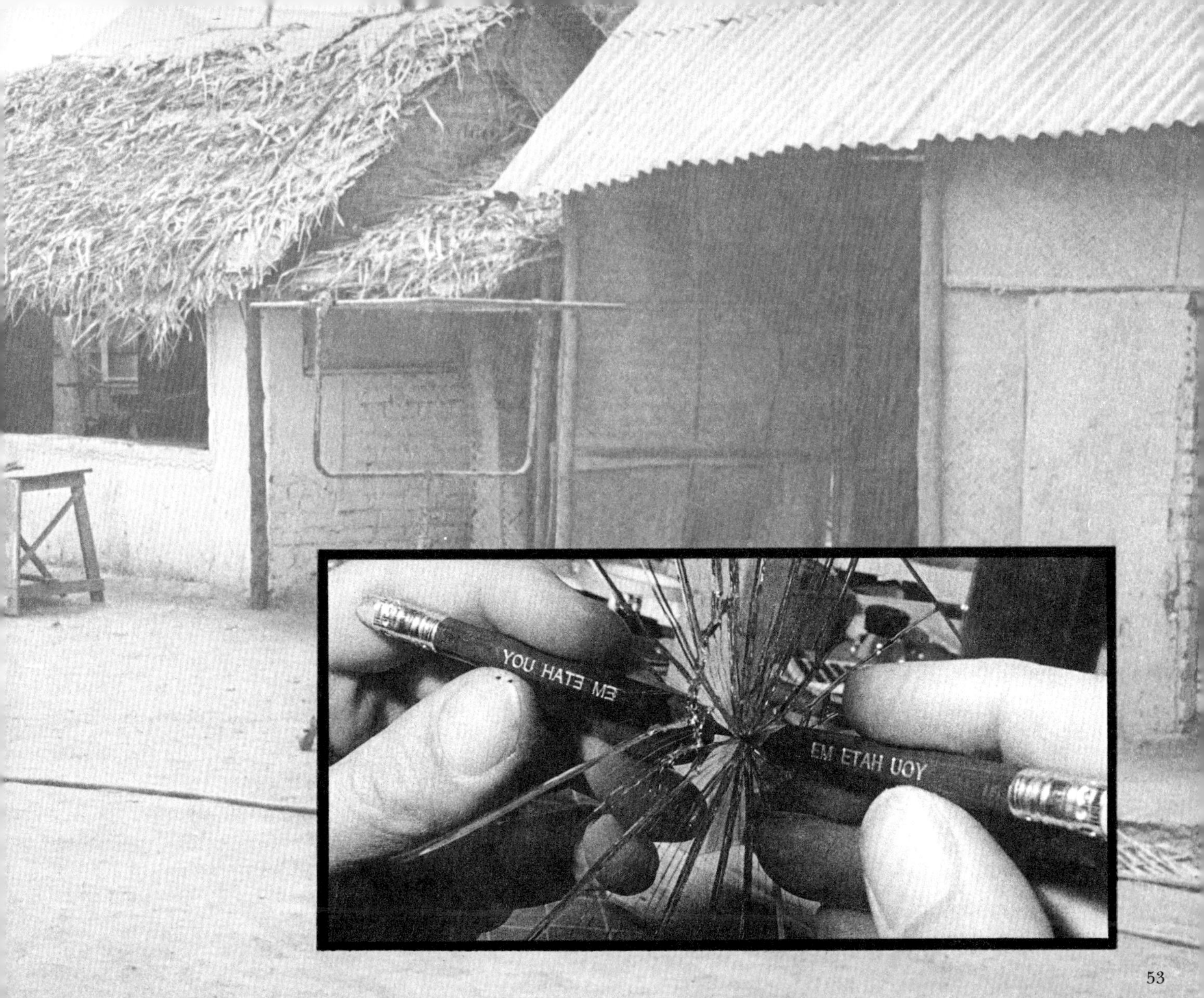
YOU HATE ME
EM ETAH UOY

UNIT 5: EXERCISES

1. Which ending is best?
2. Is Em Etah Uoy a good person?
3. Are Third World people important?
4. Do you prefer the term "Developing World" to "Third World"? If so, why?
5. What do you think of interracial couples?
6. Soap operas are: a) slippery b) clean c) full of arias d) none of the above.
7. Is science sexy? If so, why?

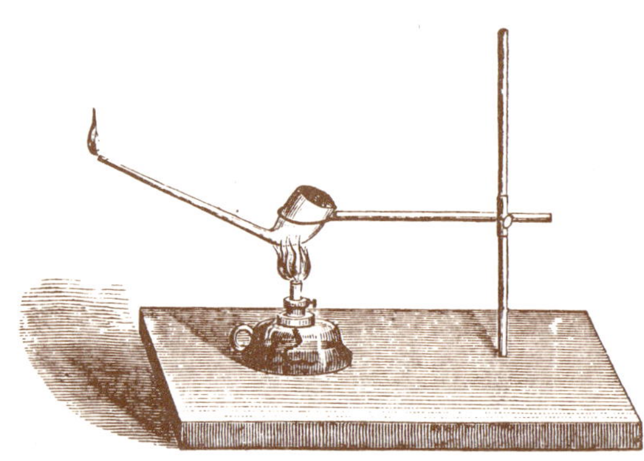

UNIT 6: THE PARTICLE THEORY OF HISTORY

Synthetic foliage used as a neutralizing foreground element.

History can be regarded as an accumulation of past decisions and deeds (particles) that form the major events recorded by scholars. Historical periods culminate in the creation of large monuments. In the interest of objectivity, historians standardize their viewpoints by the use of neutralizing elements.

1873

1982

At Cambridge University in 1964, an attempt was made to predict upcoming history through the use of a board game in which the players represented historical forces.

The experiment, however, proved to be quite dangerous and had to be discontinued.

Historians have often remarked that some of the nicest people become victims of historical forces far beyond their comprehension.

The Slaves and Their Hardships, a one act play, 1870.

UNIT 6: EXERCISES

1. Do you believe that History means anything?
2. Are you a victim of History? Do you know someone who is?
3. Is it pleasurable to view victims of historical forces?

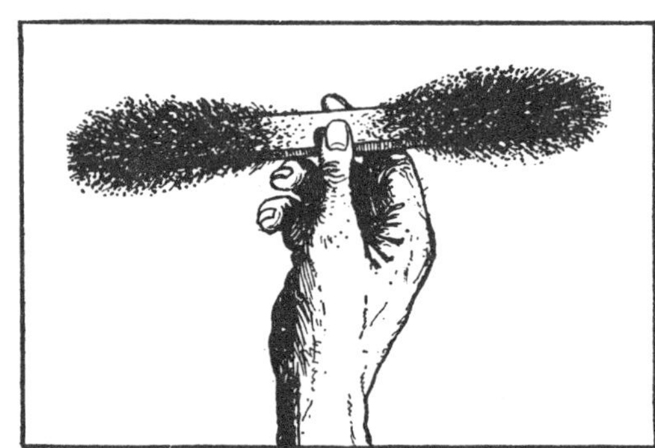

UNIT 7: PARTICLE THEORY AND MORALITY

Morality is defined as a doctrine or system relating to principles of right and wrong in behavior. The idea of responsibility for behavior was initially undermined by Freud's concept of the unconscious. Particle theorists now make computer models of human minds that explain decision making to a complex series of binary decisions.

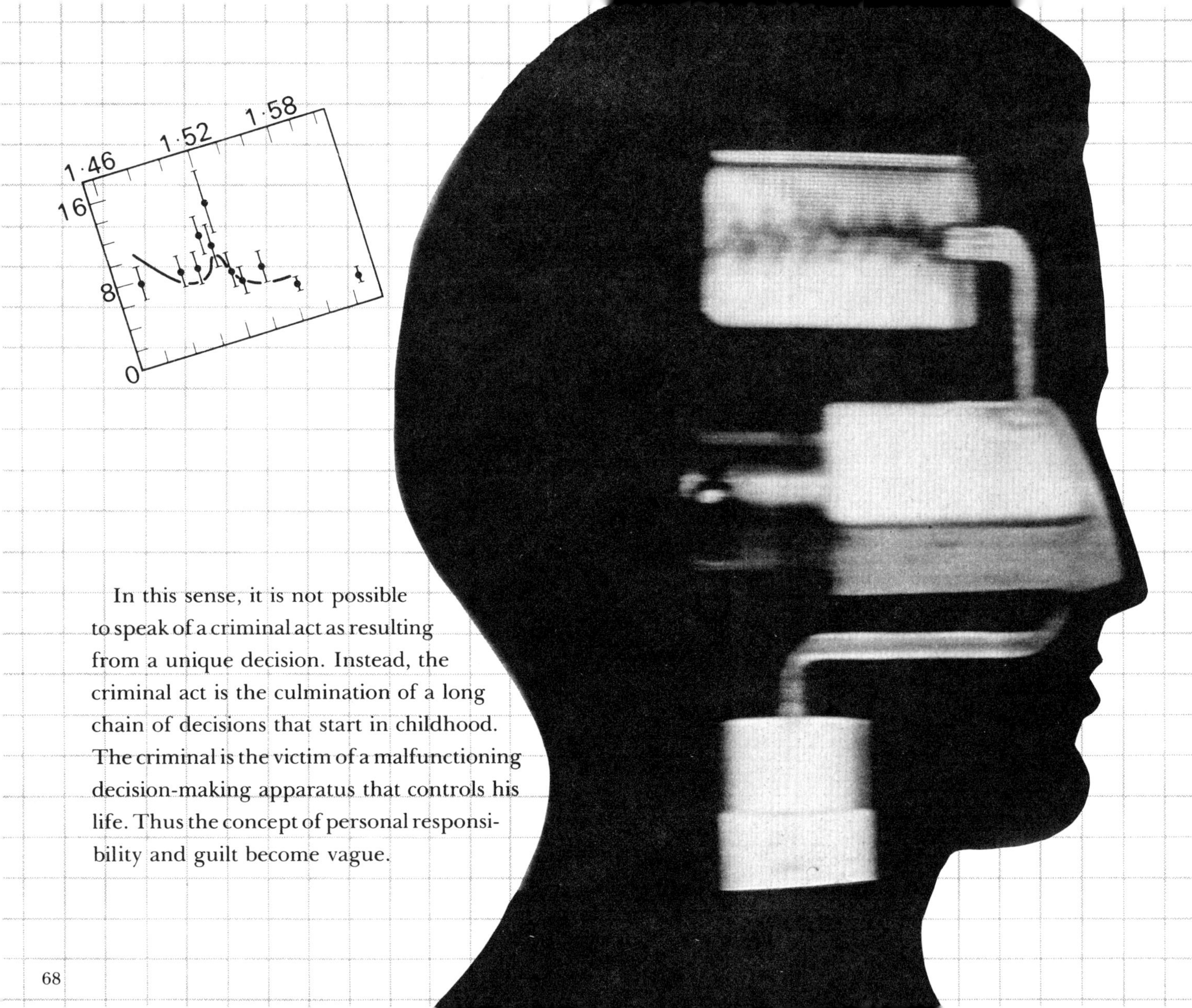

In this sense, it is not possible to speak of a criminal act as resulting from a unique decision. Instead, the criminal act is the culmination of a long chain of decisions that start in childhood. The criminal is the victim of a malfunctioning decision-making apparatus that controls his life. Thus the concept of personal responsibility and guilt become vague.

Wednesday, September 20, 1989

Police: Boyfri killed girlfrie destroyed he

NEW YORK (AP) — A spurned short-order cook was arrested Monday for allegedly killing his girlfriend, dismembering her body and then boiling her skin, police said.

"The relationship had gone sour, and she wanted other people to occupy the apartment," Fenrich said. The couple had shared the apartment.

Rakowitz allegedly beat and stabbed the woman to death and kep the body in the apartment, Fenric said.

"For approximately one week a ter the killing, the suspect continued to dismember and destroy the victim's body parts," the chief said, adding that police were investigating the possibility that cannibalism also might have occurred.

"He basically boiled the body and flushed the skin down the toilet," said Clifford.

The remains were in a five-gallon container and were partially disguised with kitty litter, said S John Clifford, another poli man.

The by th Th iden Aug rel

a ident ative So disc day Po

THE NEW YORK TIMES **METROPOLITAN** SUNDAY, AUGUST 12, 1990

Father Is Held and Confesses to Newborn's

By JAMES C. McKINLEY Jr.

A 19-year-old man who had moved to Queens from Nebraska two weeks ago was charged yesterday with killing his 6-day-old son, dismembering the body and then feeding it to a German shepherd he was training as a guard dog, the police said.

Detectives said yesterday that the baby's crying early Friday morning woke the father, Jason Radtke, who then took the infant out of a crib and began to walk him. When the baby suddenly wet him, Mr. Radtke became enraged, the police said, and threw the child to the floor, killing him.

Mr. Radtke is believed to have then dismembered the baby's body with a razor and left the remains "in a position to be consumed by the German shepherd," Capt. John Creegan of the New York detectives said.

To Cover Up the Deat

Captain Creegan said had confessed in writte ments that he threw floor and dismembere the death.

Mr. Radtke was a Precinct station night after being c tives. He was ch of second-degr

tional murd difference Mr.

The collective guilt or responsibility of a society is even more difficult to define. In ancient societies, sacrifice was a religious function. Similarly, the atom bomb, with its history of and potential for destruction, is a sacred object that may prevent war. For contemplation of the atom bomb results in images of unbelievable destruction and the conviction that war must be avoided at all cost.

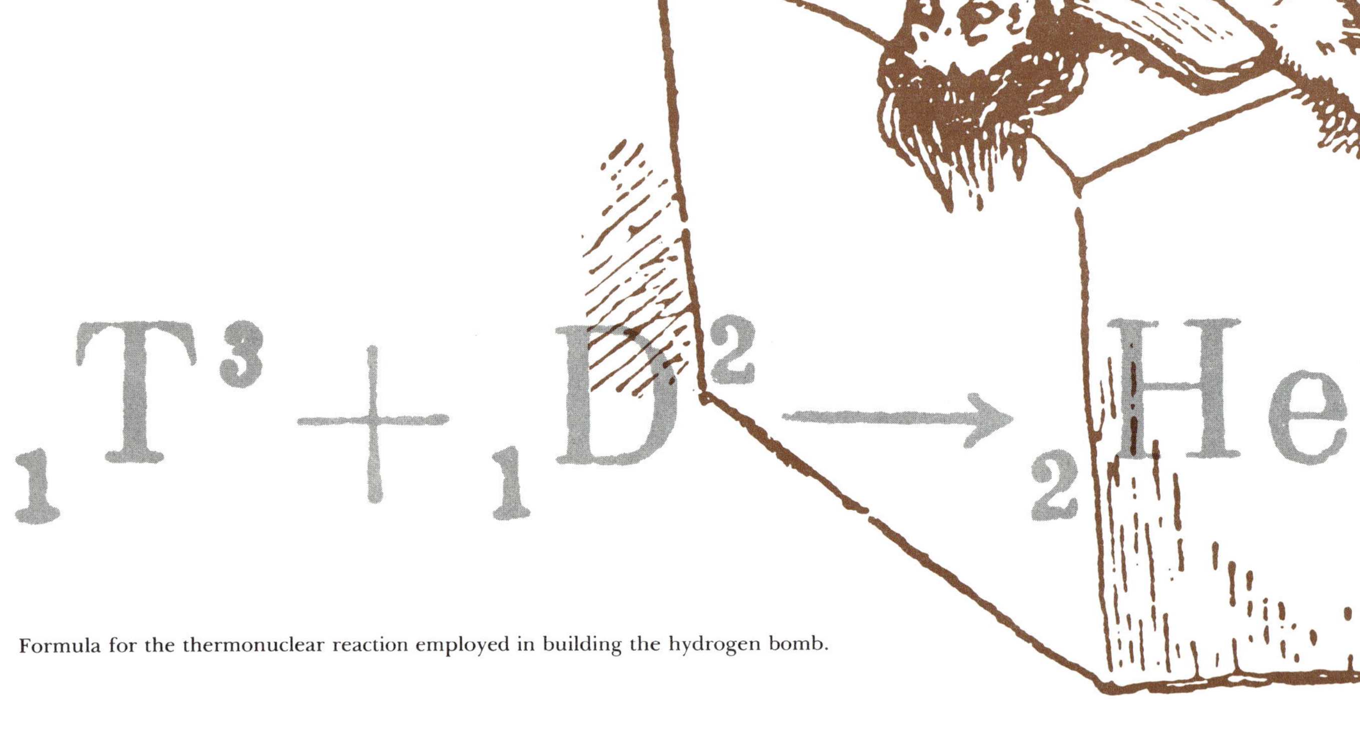

Formula for the thermonuclear reaction employed in building the hydrogen bomb.

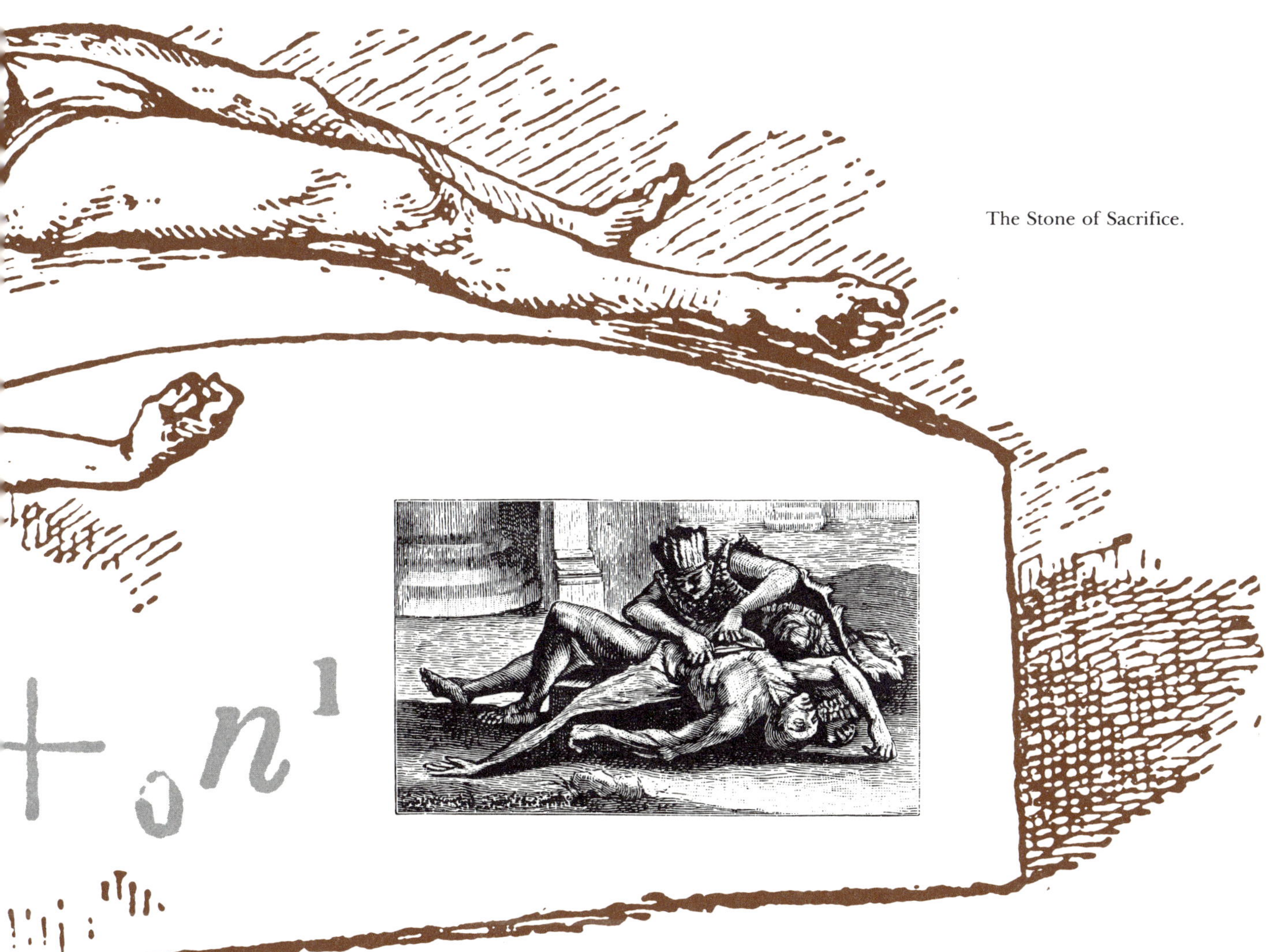

The Stone of Sacrifice.

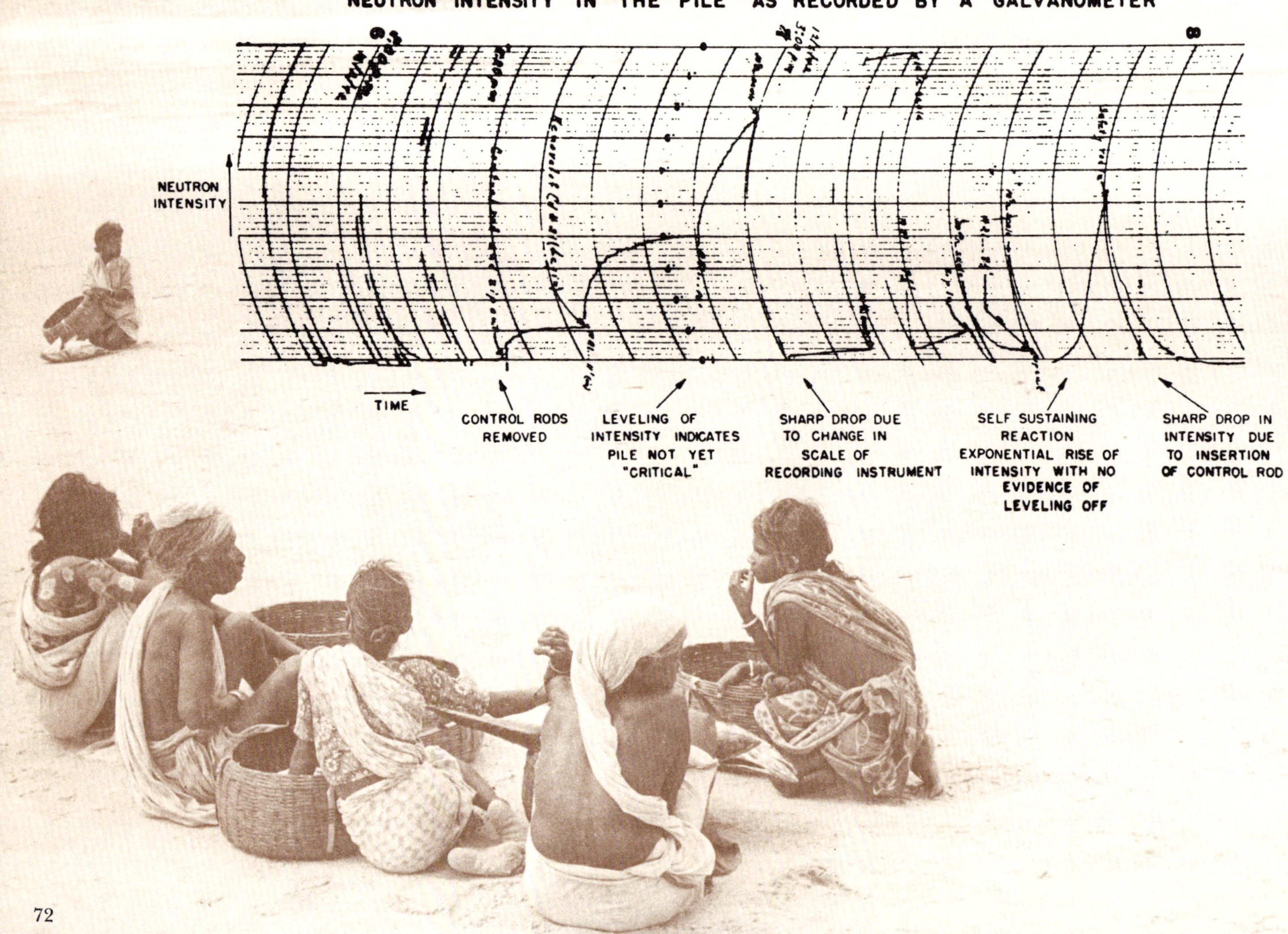
DEC. 2 1942 START-UP
OF
FIRST SELF-SUSTAINING CHAIN REACTION
NEUTRON INTENSITY IN THE PILE AS RECORDED BY A GALVANOMETER
NEUTRON INTENSITY
TIME
CONTROL RODS REMOVED
LEVELING OF INTENSITY INDICATES PILE NOT YET "CRITICAL"
SHARP DROP DUE TO CHANGE IN SCALE OF RECORDING INSTRUMENT
SELF SUSTAINING REACTION EXPONENTIAL RISE OF INTENSITY WITH NO EVIDENCE OF LEVELING OFF
SHARP DROP IN INTENSITY DUE TO INSERTION OF CONTROL ROD

The creation of the atom bomb required a deep understanding of the interaction of atomic particles. Thus Western scientific, moral, and historical particle theories culminate with this invention. It is not surprising that the great Mahatma Gandhi, when asked by a reporter what he thought of Western civilization, replied: "It would be a good idea."

UNIT 7: EXERCISES

1. What is morality?
2. Can a particle be moral?
3. Can a system of particles be moral?
4. Do you thrill to the power of the atom?
5. Is civilization moral?
6. Are answers to these questions dangerous?

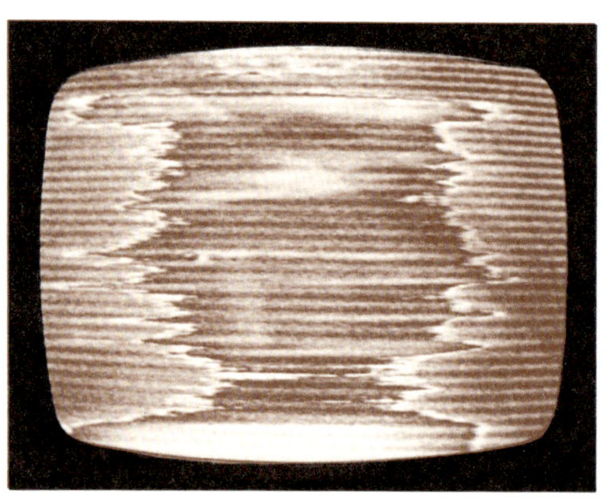

BIBLIOGRAPHY

Arrow-Brown, Florence [Bibo]. *Those Small Parts!* New York: Little Golden Books, 1989.

Blodgett, Captain Bob. *Particles Ahoy*. Boothbay, Maine: Episcopalian Press, 1847. Reprint. Portland, Maine: Ocean Books, 1978.

Dawkins, Richard. *The Selfish Gene*. New York: Oxford University Press, 1976.

Deschamps, F. *Particle Politics*. Berkeley, California: Papyrus Productions, 1985.

Duchamps, M. *La Partie Cul*. Paris: Hachette, 1921.

Hobart, Jack [Lola Galore]. *Nights of the Torrid Particles*. New York: Harlequin, 1990.

Lopez, Jesus Martinez. "Las Partículas Desnudas." *International Journal of Pataphysics*, 93 (Nov. 1985): 198-202.

Lucretius, Carus Titus. *De Rerum Natura*. Translated by Cyril Bailey. Oxford: Clarendon Press, 1949.

Mohns, J.L. "Particle Psychology: A Study in Handling Neurotic Particles," *Psychology Today*, 22 (Feb. 1989): 121-128.

Shiva, S.R.V. "Destruction By Particles," *Physical Digest*, 24 (Spring 1986): 13-22.

Vishnu, M.V.S. "Preservation of Particles: An Analysis of Opposing Viewpoints," *Physical Digest*, 24 (Fall 1986): 47-53.

"You cannot wrap fire in paper."
Chinese proverb

This project was generously supported by two grants–one from Art Matters Inc., New York, New York, and the other from Hallwalls Contemporary Arts Center, Buffalo, New York. Hallwalls' funding came from the National Endowment for the Arts and the Rockefeller Foundation.

Halftones and film work for this book were done by Brad Freeman at the Borowsky Center for Publication Arts at the University of the Arts, Philadelphia, Pennsylvania. Editing was done by Christiane Deschamps.

Thanks to Nexus Press for the Individual Artist Project and the people who made this book possible: Michael Goodman, JoAnne Paschall, Paul Trautwein, and Kim Knox.

This book was printed on Mohawk Superfine, an acid-free, archival paper manufactured by Mohawk Paper Mills, Cohoes, New York.

Nexus Press, a programming division of Nexus Contemporary Art Center, is supported in part by funds from the National Endowment for the Arts, Georgia Council for the Arts, Bureau of Cultural Affairs–City of Atlanta, and the Fulton County Arts Council–Fulton County Commission.

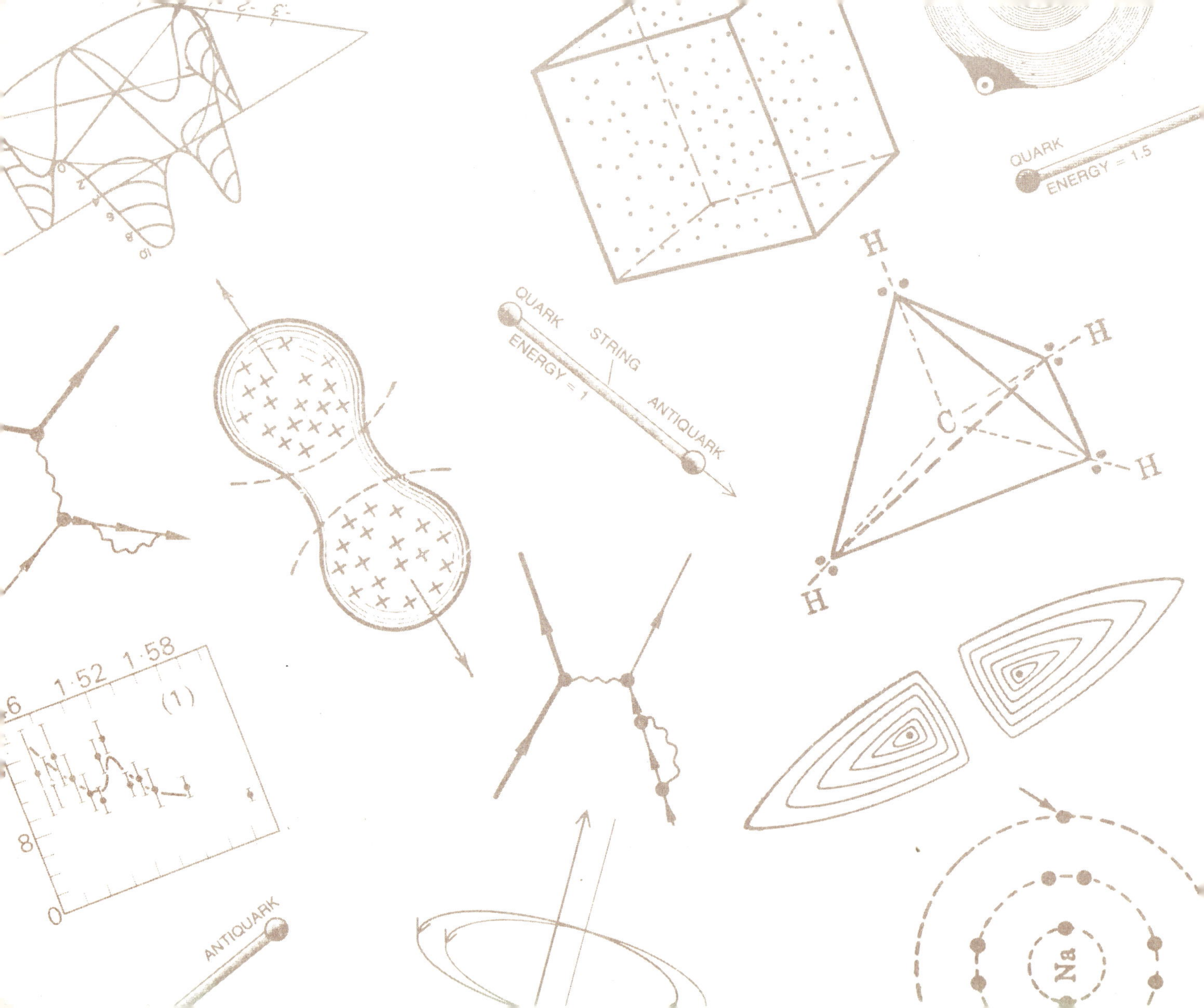

QUARK
ENERGY = 1.5
QUARK
STRING
ENERGY = 1
ANTIQUARK
H
H
C
H
H
1·52
1·58
(1)
8
0
ANTIQUARK
Na